Grades 5-8

FOCUS ON

MIDDLE SCHOOL CHEMISTRY

Teacher's Manual

3rd Edition

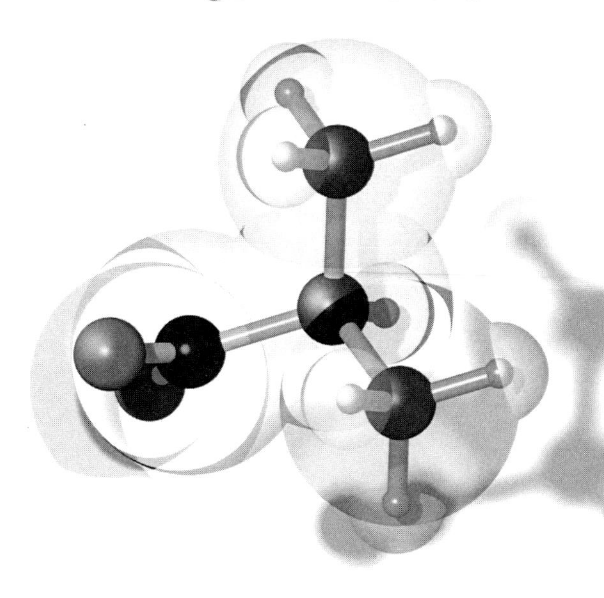

Rebecca W. Keller, PhD

Real Science-4-Kids

Cover design: David Keller
Opening page: David Keller, Rebecca W. Keller, PhD
Illustrations: Rebecca W. Keller, PhD

Focus On Middle School Chemistry Teacher's Manual—3rd Edition
ISBN 978-1-941181-53-9

Published by Gravitas Publications Inc.
www.gravitaspublications.com
www.realscience4kids.com

A Note from the Author

This curriculum is designed to engage middle school level students in further exploration of the scientific discipline of chemistry. The *Focus On Middle School Chemistry Student Textbook—3rd Edition* and the accompanying *Laboratory Notebook* together provide students with basic science concepts needed for developing a solid framework for real science investigation into chemistry.

The experiments in the *Laboratory Notebook* allow students to expand on concepts presented in the *Student Textbook* and develop the skills needed for using the scientific method. This *Teacher's Manual* will help you guide students through the laboratory experiments.

There are several sections in each chapter of the *Laboratory Notebook*. The section called *Think About It* provides questions to help students develop critical thinking skills and spark their imagination. The *Experiment* section provides students with a framework to explore concepts presented in the *Student Textbook*. In the *Conclusions* section students draw conclusions from the observations they have made during the experiment. A section called *Why?* provides a short explanation of what students may or may not have observed. And finally, in each chapter an additional experiment is presented in *Just For Fun*.

The experiments take up to 1 hour. Materials needed for each experiment are listed on the following pages and also at the beginning of each experiment.

Enjoy!

Rebecca W. Keller, PhD

Materials at a Glance

Experiment 1	Experiment 3	Experiment 5	Experiment 6	Experiment 7
imagination **Experiment 2** 10 ml glass graduated cylinder glass eyedropper 60 ml (1/4 cup) water 60 ml (1/4 cup) rubbing alcohol 60 ml (1/4 cup) vegetable oil waterproofing substance, e.g., wax, Scotch-Gard additional water and vegetable oil (small amount) **Optional** disposable glass tube Goo Gone or similar cleaner	food labels, several periodic table of elements from *Student Textbook* resources (books or online) **Optional** computer with internet access **Experiment 4** colored marshmallows, 1 pkg small, 1 pkg large (or gumdrops and/or jellybeans) toothpicks, 1 box marking pen **Optional** food coloring	baking soda lemon juice balsamic vinegar salt and water: 15-30 ml salt dissolved in 120 ml water (1-2 tbsp. salt dissolved in 1/2 cup of water) 2 or more egg whites milk small jars, 7 or more measuring cups and spoons eye dropper Peanut Brittle (see next page–Foods) pan, buttered saucepan	one head of red cabbage distilled water, about 1 liter (1 quart) various solutions, such as: ammonia vinegar clear soda pop milk mineral water large saucepan knife several small jars white coffee filters eyedropper measuring cup measuring spoons marking pen scissors ruler suggested natural materials (see next page)	red cabbage juice indicator (from Experiment 6) household ammonia vinegar large glass jar measuring spoons measuring cup household solutions chosen by students (to test for acidity and basicity)

Experiment 8	Experiment 9	Experiment 10	Experiment 11	Experiment 12
tincture of iodine [Iodine is VERY poisonous — DO NOT LET STUDENTS EAT any food items with iodine on them.] a variety of raw foods, including: pasta bread celery potato banana (ripe) and other fruits 1 unripe (green) banana liquid laundry starch (or equal parts borax and corn starch mixed in water) absorbent white paper eye dropper cookie sheet marking pen knife	about 120 ml (1/2 cup) each: water ammonia vegetable oil rubbing alcohol melted butter vinegar small jars (7 or more) food coloring (6 colors) dish soap, 30 ml (2 tbsp.) eyedropper measuring cup measuring spoons marking pen spoon ballpoint ink pens of various colors, including black (see bottom of next page)* rubbing alcohol coffee filters, several (white) cardboard shoebox (or similar size box) tape scissors ruler	1-2 brown paper bags cut into about 20 5 cm x 5 cm (2″x 2″) squares wax paper, 8 pieces paper towels tape knife scissors ruler marker or pen foods: vegetable oil butter celery potatoes banana avocado lard margarine water orange cheese milk cream several food products labeled fat free and low fat	liquid laundry starch, 120 ml (1/2 cup); or 10 ml (2 tsp.) borax and 10 ml (2 tsp.) cornstarch Elmer's white glue, 60 ml (1/4 cup) Elmer's blue glue (or another different glue), 60 ml (1/4 cup) water 2 small jars marking pen that will write on glass Popsicle sticks for stirring measuring cup safety goggles rubber gloves apron 10 ml graduated cylinder beaker or glass jar glass stirring rod Nylon Synthesis and Rope Trick Kit from Home Science Tools**	tincture of iodine [VERY POISONOUS—DO NOT LET STUDENTS EAT] bread (1-2 slices) timer wax paper marking pen cup one raw egg one raw onion table salt clear liquid dish washing detergent rubbing alcohol (isopropanol, 70-90%) wooden stir stick or Q-tip coffee filter (any color) sieve 2 glass jars or large test tubes measuring cup and measuring spoons blender

** Nylon Synthesis and Rope Trick Kit from Home Science Tools: http://www.hometrainingtools.com/, Item # KT-ISNYLON

Materials

Quantities Needed for All Experiments

Equipment

apron
beaker or glass jar
blender
cookie sheet
cup
eyedropper
eyedropper, glass
gloves, rubber
goggles, safety
graduated cylinder, glass, 10 ml
jar, glass, 2, or large test tubes
jar, large glass
jar, small, 7 or more
knife
measuring cups
measuring spoons
pan
rod, glass stirring
ruler
saucepan, large
scissors
sieve
spoon
timer

Optional

computer with internet access

Other

natural materials, (suggested) Exper. 6: poppyseed or cornflower petals; madder plant (Rubiaceae family); red beets; rose petals; berries; blue and red grapes; cherries; geranium petals; morning glory; red onion; petunia petals; hibiscus petals (or hibiscus tea); carrots; other strongly colored plant materials of students' choice
periodic table of elements from *Student Textbook*
resources (books or online)

Materials

alcohol, rubbing (isopropanol, 70-90%), at least 180 ml (3/4 cup)
ammonia, household
coffee filter (any color)
coffee filter, white, several
dish soap
dish washing detergent, liquid, clear
food coloring (6 colors)
glue, Elmer's blue (or another glue different from white), 60 ml (1/4 cup)
glue, Elmer's white, 60 ml (1/4 cup)
iodine, tincture of [VERY poisonous: DO NOT LET STUDENTS EAT any food items with iodine on them]
labels, food (student chosen), several
Nylon Synthesis and Rope Trick Kit from Home Science Tools: http://www.hometrainingtools.com/ Item # KT-ISNYLON
paper, absorbent white
paper bag, brown, 1-2
paper towels
pen, marking
pen, marking, that will write on glass
pens, ballpoint ink of various colors, including black (see below)*
Popsicle sticks for stirring
shoebox, cardboard (or similar size box)
solutions, household, chosen by students (to test for acidity and basicity)
starch, liquid laundry (or equal parts borax and corn starch mixed in water)
stick, wooden stir stick or Q-tip
tape
toothpicks, 1 box
water, distilled, about 1 liter (1 quart)
waterproofing substance, such as car wax, floor wax, silicone spray, or Scotch-Gard (small amount
wax paper

Optional

Goo Gone or similar cleaner
tube, glass, disposable

Foods

baking soda
banana, 2 unripe (green)
bread (1-2 slices)
butter, about 120 ml (1/2 cup)
cabbage, red, 1 head
cheese
cream
egg, one raw
egg whites, 2 or more
fat free food products, several
lard
lemon juice
margarine
marshmallows, colored, 1 pkg small, 1 pkg large (or gumdrops and/or jellybeans)
milk
mineral water
onion, one raw
peanut brittle ingredients
- 360 ml (1 1/2 cups) sugar
- 240 ml (1 cup) white corn syrup
- 120 ml (1/2 cup) water
- 360 ml (1 1/2 cups) raw peanuts (can be omitted)
- 5 ml (1 teaspoon) baking soda
- butter to grease pan

raw foods, including:
- avocado
- banana (ripe) and other fruits
- bread
- celery
- orange
- pasta
- potato

salt, 15-30 ml (1-2 tbsp.) or more
soda pop, clear
vegetable oil, about 180 ml (3/4 cup)
vinegar
vinegar, balsamic
water

* Experiment 9—Pens: Regular Bic® or other brand ballpoint pens can be used in this experiment. Black, blue, red, and green will give enough colors to compare. Also, multicolored ballpoint pens work well. Try to find one with at least 7 or 8 different colors. Ballpoint pens work better than felt tip pens or markers. Buy inexpensive pens for this experiment because they will be taken apart.

Contents

Experiment 1

Learning to Argue Scientifically

Materials Needed

- imagination

Objectives

In this experiment students will explore thought experiments and presenting scientific arguments.

The objectives of this lesson are for students to:

- Explore thought experiments.
- Explore the development of scientific arguments.

Experiment

I. Think About It

Read this section of the *Laboratory Notebook* with your students.

Ask questions such as the following to guide open inquiry.

- *Do you think it is important for scientists to have arguments about their theories? Why or why not?*
- *Do you think arguing helps scientists better understand science? Why or why not?*
- *Do you think it is helpful for scientists to practice arguing? Why or why not?*
- *Do you think an experiment can be performed by thinking about it? Why or why not?*
- *What do you think a scientist needs to think about before performing an experiment?*
- *What do you think a scientist might need to think about after an experiment has been completed?*

II. Experiment 1: Learning to Argue Scientifically—A Thought Experiment

A thought experiment is done by thinking scientifically about how something might work without actually doing an experiment. In this experiment students will read a fictional play to gain an understanding of how scientists argue their theories, and they will begin to learn how to form the basis of a scientific argument by thinking about it.

Have the students read the entire experiment.

Objective: An objective is provided.
Hypothesis: A hypothesis is provided.

EXPERIMENT

❶ Have the students read the play *The Mystery of Substance: A Philosophical Play* by D. R. Megill.

❷-❹ Have the students answer the questions about the play. There are no "right" answers.

Results

❶-❷ Have the students answer the questions based on what they learned from the play. There are no "right" answers.

III. Conclusions

Have the students use their observations to draw conclusions about arguing scientifically.

IV. Why?

Read this section of the *Laboratory Notebook* with your students.
Discuss any questions that might come up.

V. Just For Fun

Students are to imagine they are on a newly discovered planet that appears to be made entirely of candy. They are to think of experiments to do to prove or disprove this theory. There are no right answers to this thought experiment.

Experiment 2

Reading the Meniscus

Materials Needed

- 10 ml glass graduated cylinder
- glass eyedropper
- 60 ml (1/4 cup) water
- 60 ml (1/4 cup) rubbing alcohol
- 60 ml (1/4 cup) vegetable oil
- waterproofing substance, such as car wax, floor wax, silicone spray, or Scotch-Gard (small amount)
- additional water and vegetable oil (small amount)

Optional

- disposable glass tube
- Goo Gone or similar cleaner

Objectives

In this experiment students will explore how to correctly read a graduated cylinder.

The objectives of this lesson are for students to:

- Practice using chemistry lab equipment.
- Explore how different liquids behave in glass lab equipment.

Experiment

I. Think About It

Read this section of the *Laboratory Notebook* with your students.

Ask questions such as the following to guide open inquiry. After the students have discussed their ideas, they can experiment with placing a droplet of water and a droplet of oil on various surfaces to observe what happens.

- *If you put a droplet of water on a plastic surface, what do you think will happen?*
- *If you put a droplet of water on a glass surface, what do you think will happen?*
- *If you put a droplet of oil on a plastic surface, what do you think will happen?*
- *If you put a droplet of oil on a glass surface, what do you think will happen?*
- *Why do you think water spreads out on a glass surface and oil does not?*

II. Experiment 2: Reading the Meniscus

Have the students read the entire experiment before writing an objective and a hypothesis.

Objective: Have the students write an objective (What will they be learning?). Some examples:

- *To find out how oil and water will behave in a glass graduated cylinder.*
- *To learn how to read the volume of water in a glass graduated cylinder.*

Hypothesis: Have the students write a hypothesis (what they think they will be learning). The hypothesis can restate the objective as a statement that can be proved or disproved by their experiment. Some examples include:

- *Oil and water will behave in the same way in a glass graduated cylinder.*
- *Oil and water will behave differently in a glass graduated cylinder.*
- *It will be easy to read the volume of water in a glass graduated cylinder.*
- *It will be difficult to read the volume of water in a glass graduated cylinder.*

EXPERIMENT

❶ Have the students observe the details of the graduated cylinder. Have them note the width of the mouth, the pour spout, and the markings along the side.

❷-❸ Have the students pour 5 ml of water into the graduated cylinder and then, holding the graduated cylinder or placing it on a table, have them align their eyes with the top surface of the water. It is expected that they will notice that the water level is higher where the water is against the glass than it is in the center, creating a curve. Have the students record their observations in the chart provided in the *Results* section.

❹ Explain that the curvature of the surface of the water is called the *meniscus*. Water in a glass graduated cylinder will form a concave meniscus, with the surface of the water curving downward to the center from where the water is against the glass. Oil in a glass graduated cylinder will form a convex meniscus, with the surface of the water curving upward from the glass to the center.

Have the students observe whether the bottom of the meniscus is above or below the 5 ml mark on the graduated cylinder.

❺ Students will now need to adjust the water level until the bottom of the meniscus aligns exactly with the 5 ml mark. This may be difficult at first. It is easy to pour too much liquid in and then after pouring some out, find that too much has been poured out. Have the students use the eyedropper to add small amounts of water until the 5 ml mark aligns with the bottom of the curvature of the water.

❻-❼ Have the students repeat Steps ❷-❺ with rubbing alcohol and then with vegetable oil in that order.

Liquids with a concave meniscus are measured at the bottom of the meniscus, and liquids with a convex meniscus are measured at the top of the meniscus. In both cases the level of the liquid is being measured by its height at the center of the cylinder.

Results

A chart is provided for students to record their observations.

III. Conclusions

Have the students draw a conclusion based on their observations and research. Have them note whether their conclusion supports or does not support their hypothesis.

IV. Why?

Read this section of the *Laboratory Notebook* with your students.
Discuss any questions that might come up.

Discuss how liquids will be either attracted to or repelled by different surfaces and how this affects reading measurements in volumetric glassware, such as a graduated cylinder.

V. Just For Fun

Have the students repeat the experiment by applying a water repellent material, such as liquid car wax, floor wax, silicone spray, or Scotch-Gard, to the inside of the graduated cylinder or a disposable glass tube. To remove the water repellent from the graduated cylinder after the experiment, soak it in a hydrophobic cleaner such as Goo Gone and then wash with soap and water.

Have the students pour water into the graduated cylinder, observe the meniscus, record their results, and then repeat with vegetable oil.

A chart is provided for students to record their observations of the meniscuses formed by the liquids in the waxed cylinder and to compare the results of the *Reading the Meniscus* experiment and the *Just For Fun* experiment.

Experiment 3

What Is It Made Of?

Materials Needed

- food labels, several (students' choice)
- periodic table of elements from *Student Textbook*
- resources (books or online) such as:
  dictionary
  encyclopedia

Optional

- computer with internet access (optional)

Objectives

In this experiment students will be introduced to the concept that all things are made of atoms and will begin to explore the periodic table of elements.

The objectives of this lesson are for students to:

- Understand that atoms, or elements, are the fundamental components of all things.
- Discover that each type of atom has specific properties.

Experiment

I. Think About It

Read this section of the *Laboratory Notebook* with your students.

Ask questions such as the following to guide open inquiry.

- *Do you think all atoms are the same? Why or why not?*
- *Do you think some atoms are the same? Why or why not?*
- *Do you think it is possible to tell one atom from another? How would you do it?*
- *Do you think you can find out what properties a particular atom has? Why or why not?*
- *How would you find out what your food is made of?*
- *How would you find out what other things are made of?*

II. Experiment 3: What Is It Made Of?

Have the students read the entire experiment.

Objective: An objective is provided for this experiment:
Hypothesis: Have the students write a hypothesis. Some examples:

- *Food labels can be used to tell what is in food.*
- *I can find out what things are made of.*
- *I can use the periodic table of elements to tell me about atoms.*

EXPERIMENT

❶ Answers to the questions:

A. Protons in aluminum: 13
Electrons in aluminum: 13

B. Symbol for carbon: C

C. The elements that have chemical properties similar to helium are neon, argon, krypton, xenon, and radon.

Elements that have the same chemical properties as helium are in the same column in the periodic table.

D. Atomic weight of nitrogen: 14.0067
Number of neutrons in nitrogen: 7

❷ A table is provided for students to record information they discover about the makeup of items of their choice.

The goals of this experiment are to help students begin to investigate the things in their world and to have them start to examine what those things are made of.

There are many possible answers for this experiment. Students will begin to think about what substances are made of and how they are produced. By using basic resources such as the dictionary or encyclopedia, they may not be able to find the elemental composition of all the items they think of.

Some examples of answers are the following:

Things made of metals:

- soda cans and aluminum foil - aluminum
- silverware (steel) - iron, nickel, silver
- coins - copper, nickel
- jewelry - gold, silver

Things we eat:

- salt - sodium and chlorine
- sugar - carbon, oxygen, hydrogen
- water - hydrogen and oxygen
- bread (carbohydrates) - carbon, oxygen, hydrogen, other proteins, and other substances

Also, students can select food items with labels, such as cake mixes, cereal, noodles, and vitamins (with vitamins the label is very detailed so students can also find out how much of something is in the vitamin).

Students DO NOT need to find every component for each item. To say that a cake mix contains salt, flour, and sugar is enough. Let the students go as far as they want to with a particular item. Also, it is not necessary to look up components for each item the students think of. Have them pick a few items they are interested in researching and go from there.

Some examples of information that may be gathered:

ITEM	COMPOSITION	SOURCE
graham crackers	sodium bicarbonate (sodium)	food label
graham crackers	salt (sodium, chlorine)	food label, dictionary - page 1600
car tires	rubber (carbon and hydrogen)	Wikipedia (or www.wikipedia.org)

Results

Students will describe what they discovered about the composition of the items they researched.

Help the students write accurate statements about the data they have collected. Some examples:

- Kellogg's Sugar Smacks™ cereal contains vitamin C, which is called sodium ascorbate.
- Table salt is made of sodium and chlorine.
- Iodized table salt contains sodium, chlorine, and iodine.
- Chocolate cake mix contains sugar.
- Sugar has oxygen, hydrogen, and carbon in it.

Next, help the students think specifically about what their data show. This is an important critical thinking step that will help them evaluate future experiments.

III. Conclusions

Have the students review the results they recorded for the experiment. Have them draw conclusions based on the data they collected.

Help them write concluding statements that are valid. Encourage them to avoid stating opinions or any conclusions that cannot be drawn strictly from their data.

For example, it may be true that all cereals contain salt. However, this particular investigation cannot confirm or deny that conclusion. The most that can be stated from this investigation is "Brand X contains salt and Brand Y contains salt," but any further statement is conjecture.

Help them formulate their conclusions using the words some, all, many, and none. Point out that the statement, "All cereals contain salt," is not valid, but based on this investigation, it is valid to say, "Some cereals contain salt."

Again, there are numerous possible answers. One student may list "sugar" as a component in soup, and another may list "salt," and both answers could be correct. The true test is whether the statements about the data are valid or not valid.

Also, try to show students where broad statements can be made validly. For example, "All recent U.S. pennies contain copper" is probably a valid statement even though we haven't checked every U.S. penny.

This may seem fairly subtle, but the main point is to help them understand the kinds of valid conclusions science can offer based on scientific investigation.

IV. Why?

Read this section of the *Laboratory Notebook* with your students.
Discuss any questions that might come up.

V. Just For Fun

Students are to select one item from their list and do research to find out as much as they can about how it was made, where it was made, and where the different components might have come from.

Experiment 4

Modeling Molecules

Materials Needed

- small, colored marshmallows, 1 pkg
- large marshmallows, 1 pkg (could also use gumdrops and/or jellybeans in place of marshmallows)
- toothpicks, 1 box
- marking pen

Optional

- food coloring

Objectives

In this experiment students will explore how atoms combine to make molecules.

The objectives of this lesson are for students to:

- Observe how making models is helpful in understanding how atoms combine to form molecules.
- Explore the concept that atoms follow rules when combining to make molecules.

Experiment

I. Think About It

Read this section of the *Laboratory Notebook* with your students.

Ask questions such as the following to guide open inquiry.

- *Do you think it's important for people to have rules to follow? Why or why not?*
- *What do you think life would be like if people did not have any rules?*
- *Do you think atoms have rules to follow when they combine to make molecules? Why or why not?*
- *What do you think life on Earth would be like if atoms could combine in any way they wanted to with no rules?*

II. Experiment 4: Modeling Molecules

In this experiment students will use marshmallows and toothpicks to explore the ways in which atoms combine to form molecules.

Have the students read the entire experiment.

Objective: An objective is provided.
Hypothesis: Have the students write a hypothesis. Some examples:

- *Models can show how atoms combine.*
- *By making models, I can see how atoms follow rules.*

EXPERIMENT

❶-❷ Two sizes of marshmallows are preferred for this experiment. Gumdrops and jellybeans can also be used.

Have the students make marshmallow and toothpick molecules without following any rules. Encourage the students to make molecules of various sizes and shapes. They do not need to record every shape they make, but have them draw at least several different shapes.

All of their answers will be correct since all shapes are valid in this step. Encourage them to use their imagination in combining the marshmallows.

❸ Students will make "real" molecule models following specific rules.

The rules for carbon, nitrogen, oxygen, hydrogen, and chlorine are shown. Note that the orientation of the bonds (toothpicks) is also important. Before making molecules, the students can first practice putting the toothpicks into several marshmallows while following these rules.

Note that the large marshmallows are assigned to the atoms carbon, nitrogen and oxygen. If this is confusing, try to differentiate between the atoms by color-coding each with a drop of food coloring.

Results

❶ Now the students will follow the rules and make "molecules" with the marshmallow "atoms." Have them draw their molecules.

These illustrations show the correct shapes for the molecules that are given. (Drawings may vary.)

Have the students note the number of bonds for each molecule, and ask them whether or not they followed the rules.

(Drawings may vary.)

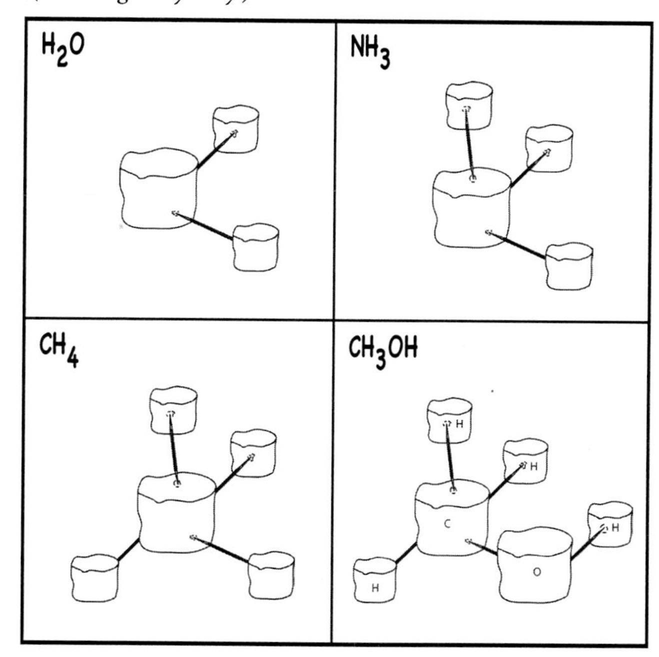

III. Conclusions

Have the students write some conclusions about the molecules they have created with the marshmallows. Help them try to be specific with the conclusions they write.

IV. Why?

Read this section of the *Laboratory Notebook* with your students.
Discuss any questions that might come up.

V. Just For Fun

Students will follow the rules and make their own "molecules." For each molecule model, have the students note how many bonds each "atom" forms and how many of each type of atom are in the molecule.

Have them draw their molecule models.

Some suggested molecules are the following:

CCl_4 four chlorine atoms attached to one central carbon (carbon tetrachloride)

CH_3CH_3 two carbon atoms connected to each other with three hydrogens each (ethane)

CH_3CH_2OH two carbon atoms connected to each other. One carbon atom has three hydrogens attached. The other has two hydrogens and an oxygen attached to it, and the oxygen has a hydrogen attached. (ethanol)

The students can build many different molecules while still following the rules.

(Answers will vary. Some examples are shown.)

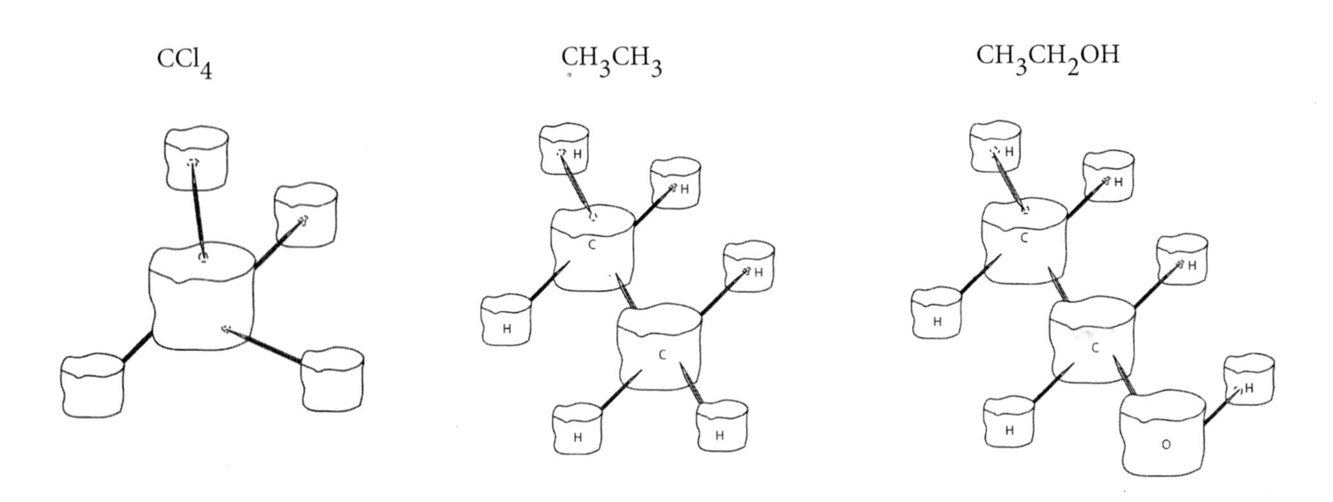

Experiment 5

Identifying Chemical Reactions

Materials Needed

- baking soda
- lemon juice
- balsamic vinegar
- salt and water:
  15–30 ml salt dissolved in
  120 ml water
  (1–2 tbsp. salt dissolved in
  1/2 cup of water)
- 2 or more egg whites
- milk
- small jars, 7 or more
- measuring cups and spoons
- eye dropper

Peanut Brittle

- 360 ml (1 1/2 cups) sugar
- 240 ml (1 cup) white corn syrup
- 120 ml (1/2 cup) water
- 360 ml (1 1/2 cups) raw peanuts (can be omitted)
- 5 ml (1 teaspoon) baking soda
- buttered pan

Objectives

In this experiment students will observe how chemical reactions cause bonds to form and break, creating changes that can be seen.

The objectives of this lesson are for students to:

- Observe evidence that chemical reactions are taking place.
- Chart their observations.

Experiment

I. Think About It

Read this section of the *Laboratory Notebook* with your students.

Ask questions such as the following to guide open inquiry.

- *Do you think chemical reactions are important in your daily life? Why or why not?*
- *Do you think any two liquids mixed together will undergo a chemical reaction? Why or why not?*
- *If you mix two liquids together and they have a chemical reaction, do you think there will be a chemical reaction every time more of those liquids are mixed? Why or why not?*
- *Do you think you can tell whether or not a chemical reaction has happened? Why or why not?*
- *What clues would you look for to show that a chemical reaction has taken place?*
- *What clues would you look for to show that a chemical reaction has not taken place?*

II. Experiment 5: Identifying Chemical Reactions

In this experiment students will examine chemical reactions and try to identify when they occur. Balsamic vinegar is recommended because the reaction when it is mixed with baking soda will be more dramatic, but other kinds of vinegar may be used.

Have the students read the entire experiment.

Objective: An objective has been provided.
Hypothesis: A hypothesis has been provided

EXPERIMENT

❶ Have the students put a small amount of each substance listed into its own jar, and then have them examine the contents of each jar, taking note of the properties of each substance. Have them record the color, texture, and odor next to each item on the materials list. For example:

- Baking Soda: white powder, no odor.
- Balsamic Vinegar: dark liquid, sour odor.

Although most of the items are food items, **do not allow the students to taste them** since tasting is not part of this experiment.

A variety of food items in addition to those on the materials list may be used. Bleach and ammonia cause good chemical reactions, but they can give off strong odors and so these chemicals are NOT RECOMMENDED.

❷ Have the students write the "reagents" (chemicals used in chemical reactions) on the top and side of the grid provided.

Have them mix some of each of two substances together in a clean jar and observe the results. Have them rinse the mixing jar in between tests. It might be interesting to the students to check the order of addition—e.g., add lemon juice to baking soda, then add baking soda to lemon juice—to see if a difference can be observed (the order of addition should not matter). This is optional.

❸ Have the students record their observations in the appropriate box for each reaction.

Results

(Expected results)

	milk	*lemon juice*	*salt water*	*baking soda*	*balsamic vinegar*	*egg whites*
milk		*REACT precipitate*	*NO*	*NO*	*REACT precipitate*	*NO*
lemon juice			*NO*	*REACT precipitate*	*NO*	*REACT precipitate*
salt water				*NO*	*NO*	*NO*
baking soda					*REACT precipitate*	*NO*
balsamic vinegar						*REACT precipitate*
egg whites						

Results for Unknown Solutions

1. Give the students two "unknown" solutions—ones that were used in the experiment but that you are not now identifying. They can either be two substances that will react or two that won't react. Have students describe the properties of each unknown.

   This part of the experiment can be done more than once. In addition, you can have the students give you "unknowns" to see if you can identify them.

   The students have observed all of the reactants both before and after a reaction. They now have the necessary knowledge to identify an unknown.

2. Have the students mix the two substances and describe the results, including why they think a chemical reaction did or did not take place.

3. Have the students record what they think the two substances are and how they identified them.

An option for additional experimentation is to give the students only one unknown. Have them guess what it might be before performing any tests. Then have the students test this unknown with each of the other reactants. Have them prove the identity of the unknown with the chemical reactions they have already observed.

III. Conclusions

Have the students review the results they recorded for the experiment and write valid conclusions. Help them state conclusions that reflect only the data found in this experiment. For example, "Salt water does not react with anything" is not a valid conclusion because we haven't tested all substances. However, "Salt water does not react with any of the items we tested" is valid.

IV. Why?

Read this section of the *Laboratory Notebook* with your students.
Discuss any questions that might come up.

V. Just For Fun

Help the students make peanut brittle following the directions provided. They will be able to observe baking soda decompose (undergo a decomposition reaction), giving off carbon dioxide gas while the peanut brittle is being made.

Peanut Brittle

360 ml (1 1/2 cups) sugar
240 ml (1 cup) white corn syrup
120 ml (1/2 cup) water
360 ml (1 1/2 cups) raw peanuts
5 ml (1 teaspoon) baking soda
buttered pan

Boil sugar, water, and syrup in a sauce pan over medium heat until the mixture turns a little brown. Add 360 ml (1 1/2 cups) raw peanuts Stir until golden brown. Don't over-brown. Add 5 ml (1 teaspoon) baking soda. Spread on buttered pan.

Have students make observations about the evidence of a chemical reaction occurring. Have them think about what other chemical reactions they have noticed when food items are being cooked or after they are cooked.

Have them record their observations and ideas in the space provided. There are no "right" answers.

Experiment 6

Making an Acid-Base Indicator

Materials Needed

- one head of red cabbage
- distilled water, about 1 liter (1 quart)
- various solutions, such as:
  ammonia
  vinegar
  clear soda pop
  milk
  mineral water
- large saucepan
- knife
- several small jars
- white coffee filters
- eyedropper
- measuring cup
- measuring spoons
- marking pen
- scissors
- ruler

Suggested Natural Materials for Just For Fun (help students select several)

- Turmeric
- Poppyseed or cornflower petals
- Madder plant (Rubiaceae family)
- Red beets
- Rose petals
- Berries
- Blue and red grapes
- Cherries
- Geranium petals
- Morning glory
- Red onion
- Petunia petals
- Hibiscus petals (or hibiscus tea)
- Carrots
- Other strongly colored plant materials of students' choice

Objectives

In this experiment students will be introduced to the concepts of acids, bases, pH, and pH indicators.

The objectives of this lesson are for students to:

- Observe that acids and bases have different properties that can be tested for.
- Use controls in an experiment.

Experiment

I. Think About It

Read this section of the *Laboratory Notebook* with your students.

Ask questions such as the following to guide open inquiry.

- *What do you think an acid is?*
- *What liquids can you think of that are acids?*
- *What do you think a base is?*
- *What liquids can you think of that are bases?*
- *How would you find out if a solution is an acid or a base?*
- *Do you think you use acids and bases in your everyday life? Why or why not?*

II. Experiment 6: Making an Acid-Base Indicator

Have the students read the entire experiment before writing an objective and a hypothesis.

Objective: Have the students write an objective (What will they be learning?). An example:

- *We will make an acid-base indicator from red cabbage and use it to determine whether solutions are acidic or basic.*

Hypothesis: Have the students write a hypothesis. An example:

- *We can use an indicator to identify solutions as acidic or basic.*

EXPERIMENT

In this experiment the students will use "controls." A control is an experiment where the outcome is already known or where a given outcome can be determined. The control provides a point of reference or comparison for experiments that use unknowns. For example, in this experiment the students will test for acidity or basicity with a pH indicator, but they do not know what the expected color change will be. By doing controls with solutions that they know are either acidic (vinegar) or basic (ammonia), they can determine what the color change for an acid is and what the color change for a base is. Only then can they test the "unknown" solutions.

The liquids in the materials list include both acids and bases. Milk in neutral. Have the students be careful when handling ammonia. Other suggested items to test include:

- water (neutral)
- Windex or other glass cleaner (basic)
- Lemon juice or orange juice (acidic)
- White grape juice (acidic)

❶-❸ Have the students prepare red cabbage juice indicator. Have them cut a head of red cabbage into several pieces, put the pieces in about 710 ml (3 cups) of boiling distilled water, and boil the cabbage for several minutes until the liquid is a deep purple color. Have them remove the cabbage and let the water cool.

❹ Have the students set aside 236 ml (1 cup) of the red cabbage juice and refrigerate the rest to use in the next experiment. It is important to refrigerate the cabbage juice or it will spoil and cannot be used for the next experiment. It should keep about two weeks in the refrigerator.

❺ Have the students cut 20 or more strips of coffee filter paper that are about 2 cm (3/4 in.) by 4 cm (1½ in.) for testing the solutions.

❻ Students will make pH paper by using an eyedropper to put several drops of the red cabbage juice on each strip of coffee filter paper and letting it dry. If the cabbage indicator is added to the strips of paper several times and dried in between, the color change when testing the liquids will be more dramatic.

❼ Have the students label one jar *Control Acid*. They will make the control acid by putting in the jar 15 ml (1 tbsp.) of vinegar and 75 ml (5 tbsp.) of distilled water.

A second jar will be labeled *Control Base*. Students will make the control base by putting in the jar 15 ml (1 tbsp.) of ammonia and 75 ml (5 tbsp.) of distilled water.

Have the students label a jar for each of the solutions they will be testing and then put 15 ml (1 tbsp.) of each substance in the appropriate jar along with 30-75 ml (2-5 tbsp.) of distilled water.

❽-❾ Students will test the *Control Acid* and the *Control Base* by dipping an unused strip of pH paper into each. Have them record their results and tape the pH papers into their *Laboratory Notebook* in the chart in the *Results* section.

The vinegar (control acid) should turn the paper pink.
The ammonia (control base) should turn the paper green.

The color change of the pH paper may be quick and noticeable or it may be subtle. It is best to have the students look at the paper immediately after it has been dipped into the solution. If it is too difficult to determine the color change of the paper, the cabbage indicator can be used directly in the solution. Have the students pour a small amount (5-10 ml [1-2 teaspoons]) into the solution and record the color change.

❿ Have the students test the remaining solutions and record their results.

III. Conclusions

Have the students review the results they recorded for this experiment. Have them draw conclusions based on the data they collected. Help the students to be specific and to make valid conclusions from their data. If a solution did not change color, but the experimental controls worked, it is probably true that the solution is neutral or near neutral. However, if no color change is observed or if the result is ambiguous, it may not be true that the solution is neutral, and it may be true that the color change is too subtle to be easily perceived.

Have the students draw conclusions even if they experienced difficulties with the experiment.

IV. Why?

Read this section of the *Laboratory Notebook* with your students.
Discuss any questions that might come up.

V. Just For Fun

Students will test different natural materials to see if they are acid-base indicators.

Help the students select and gather the materials to be tested. Have them use several materials from the list provided, and they can also try other natural materials that have a strong color. Students can use their control acid and base or other solutions that they have identified as acidic or basic.

Have the students crush (or chop) the material to be tested and put some of it in each of two small jars. They then will add an acid to one jar and a base to the other and note whether there is a color change. If there is no color change, you can direct them to make a stronger solution of the acid and base they have chosen and see if this makes a difference. They can also experiment with adding a little distilled water to the material being tested before adding the acid or base. In the chart provided, have the students record their results including whether or not they think the material is an acid-base indicator.

Some examples:

- Turmeric powder will turn red in a base (will be yellow at pH 7.4 and red at pH 8.6).
- Poppyseed or cornflower petals contain the same chemical as red cabbage and will undergo a similar color change.
- Madder plant (Rubiaceae) will turn from yellow to red in a weak base (yellow at pH 5.5 and red at pH 6.8).
- Cherries and cherry juice turn from red to purple in a base.

Additional Notes

An acid-base reaction is a type of exchange reaction. In the example illustrated in this chapter of the textbook, the molecules are not drawn with the bonds showing, and on first inspection it appears that the central carbon of both molecules has broken the rule of "4 bonds for carbon." Also, two of the oxygens appear to have broken the rule for "2 bonds for oxygen." However, in each case the bond between the central carbon atom and one of the oxygen atoms is a double bond. Double bonds are beyond the scope of this level, but all of the bonding rules are satisfied.

pH is actually a measure of the hydrogen ion concentration (written as [H]).The pH scale is important, but mathematically and conceptually the actual definition of pH is too difficult for this level. (The mathematical expression for pH is: pH = -log [H])

The higher the hydrogen ion concentration, the lower the pH; the lower the hydrogen ion concentration, the higher the pH. The hydrogen ion concentration is the real definition of what is meant by "acid" in this chapter.

Scientists measure pH with pH meters, pH paper, or solution indicators, with the use of the pH meter being the most common laboratory technique. There are a variety of pH meters and electrodes available. The most common electrode is called a glass electrode. There is a small glass ball at the end of this electrode that senses the pH electrically.

Before pH meters, pH paper was the most common way to measure pH. Litmus paper can still be found in most laboratories along with other types of pH paper. Litmus paper is made with a compound called an indicator. An indicator is any molecule that changes color as a result of a pH change.

There are two types of litmus paper—blue litmus paper tests for acidic solutions, and red litmus paper tests for basic solutions. Litmus paper is not suitable for determining the exact pH; it can only indicate whether a solution is acidic or basic. Other types of pH paper can more accurately determine the actual pH.

The chart in this section of the *Student Textbook* shows some common indicators used in the laboratory and is meant to illustrate that there are a variety of pH indicators that can be used over a wide range of pH. Often pH indicators are mixed so that more than one pH range can be detected. The names of some of these indicators are difficult to pronounce, but many of them can be looked up in a dictionary, encyclopedia, or online for pronunciation guidance.

Experiment 7

Vinegar and Ammonia in the Balance: An Introduction to Titration

Materials Needed

- red cabbage juice indicator (from Experiment 6)
- household ammonia
- vinegar
- large glass jar
- measuring spoons
- measuring cup
- household solutions chosen by students (to test for acidity and basicity)

Objectives

In this experiment students will explore acid-base reactions by performing a titration.

The objectives of this lesson are for students to:

- Perform a titration.
- Plot and analyze data on a graph.

Experiment

I. Think About It

Read this section of the *Laboratory Notebook* with your students.

Ask questions such as the following to guide open inquiry.

- *What do you think happens when you mix an acid and a base together? Why?*
- *What products do you think you would get by mixing an acid and a base together?*
- *Do you think knowing the concentration of an acid or a base could be helpful? Why or why not?*
- *Do you think there are times when you would want to have a solution that is neutral? Why or why not?*
- *Do you think plotting data on a graph can be useful? Why or why not?*

II. Experiment 7: Vinegar and Ammonia in the Balance: An Introduction to Titration

Have the students read the entire experiment before writing an objective and a hypothesis.

Objective: Have the students write an objective (What will they be learning?). For example:

- *To determine how much ammonia is needed to change the color of red cabbage juice indicator in vinegar from red to green*

Hypothesis: Have the students write a hypothesis. For example:

- *An indicator can be used to observe the acid-base reaction of vinegar and ammonia.*

EXPERIMENT

In this experiment, students will perform an acid-base titration using a red cabbage juice indicator. The red cabbage juice indicator from Experiment 6 is required. If the indicator is too old (more than two weeks or has mold or bacteria growing in it), have the students make fresh cabbage indicator.

NOTE: This titration can be tricky if the concentration of the base is too dilute.

A quick test can be performed by the teacher without the students' observation. Take 60 ml (1/4 cup) of vinegar and add indicator to it. See that it turns red. Add 60 ml (1/4 cup) ammonia directly to this acid-indicator mixture. The color should turn green, but if the color is still red, add another 60 ml (1/4 cup) of ammonia. It should turn green; however, if it does not, dilute the vinegar with 120 ml (1/2 cup) water and repeat the above steps.

This quick "titration" will help determine how much total ammonia is needed to neutralize the acid. Adjust the titration so that not much more than 60 ml (1/4 cup) of ammonia is needed. Less is all right, but the students will get frustrated if they have to add more than 100 ml (20 teaspoons) of ammonia, and the best part of the titration is the last part when they see the color change occur.

❶-❷ Have the students measure 60 ml (1/4 cup) of vinegar into a jar and add enough red cabbage juice to get a deep red color.

❸-❻ Students will add ammonia to the vinegar 5 ml (1 tsp.) at a time. Each time the students add ammonia, they will swirl the solution and then record the color of the solution and the total amount of ammonia that has been added. There is a chart provided in the *Results* section.

As the students add ammonia, the color stays mostly red, then turns a little purple, and finally turns all green. The transition is quite striking. Have the students continue adding ammonia to see that the color stays green.

Graphing Your Data

❶-❸ Have the students take the data from their chart and plot it on the graph provided. When all the data has been plotted, have them connect the data points.

The data points should look something like those shown in the following graph. Many points lie along the bottom left of the plot, then one or two points will be in the middle. Finally, several will be along the top right-hand side of the plot.

Have the students connect the points with a smooth curved line. Their plot should look similar to the following. Discuss the following parts of the graph:

- In the left-hand lower portion of the plot, the solution is acidic.
- In the middle portion, where the line is going upward, the solution is between acidic and basic (near neutral).
- In the upper right-hand corner, the solution is basic.

Point out that we know this because the color of the indicator is known at various pH values, as we observed in Experiment 6.

Example *(Answers may vary.)*

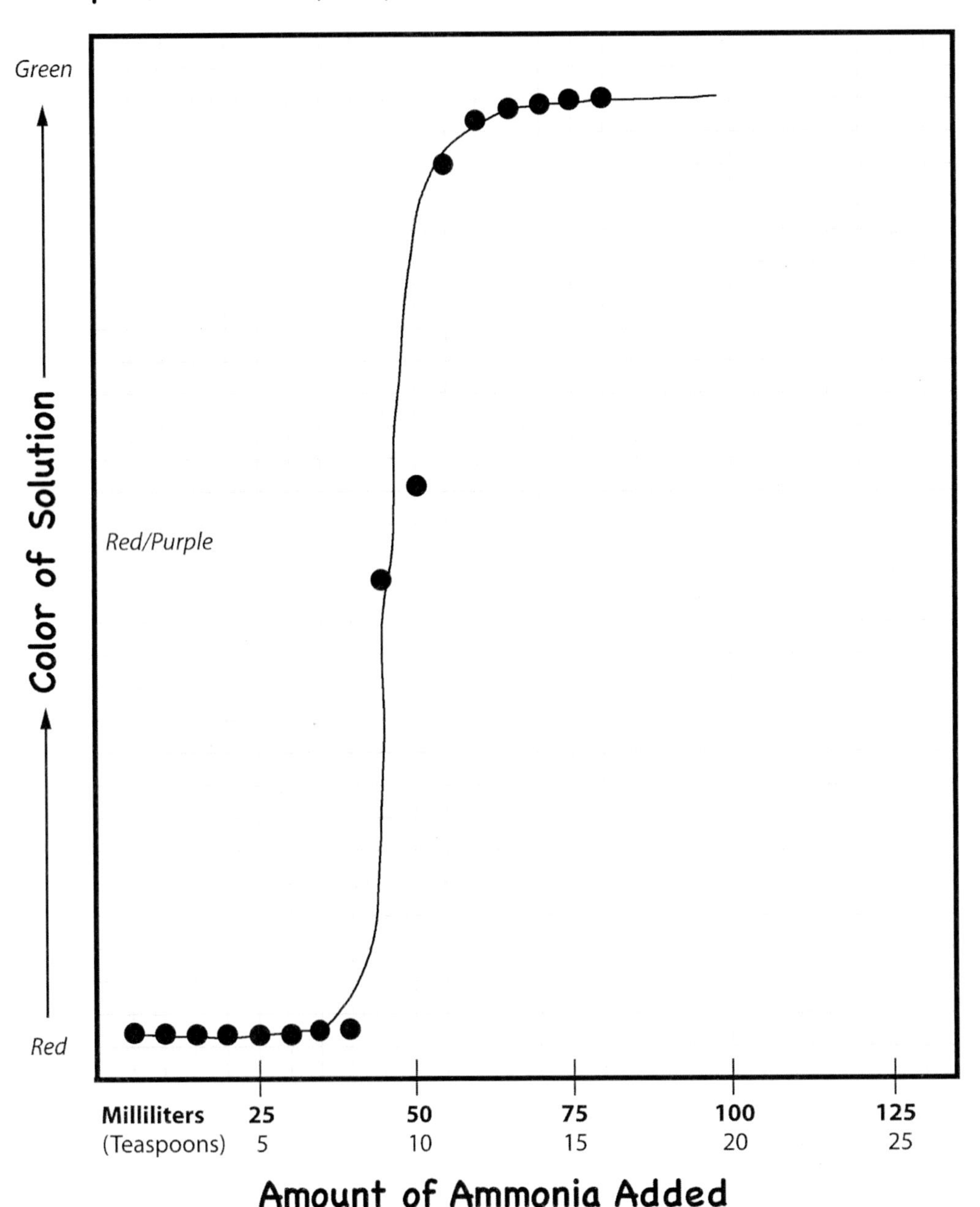

III. Conclusions

Have the students review the results they recorded for the experiment. Have them draw conclusions based on the data they collected.

Have the students write down the amount of ammonia it took to neutralize the vinegar. Note whether this equals the amount of vinegar that was used—60 ml (1/4 cup [12 tsp.]). Depending on the brands of vinegar and ammonia used, the amounts are often equal.

Help them reach valid conclusions. Some examples:

- *It took 55 ml of ammonia to turn the solution green.*
- *It took 11 teaspoons to neutralize the vinegar with ammonia.*
- *The amount of ammonia required to neutralize the vinegar was equal to the amount of vinegar used.*

IV. Why?

Read this section of the *Laboratory Notebook* with your students.
Discuss any questions that might come up.

V. Just For Fun

Students are to repeat the experiment with a different acid and base.

Help the students find household solutions that they think may be acidic or basic. Have them use red cabbage juice indicator to test each solution and then select one acid and one base. Have the students follow the steps of the previous experiment to do a titration.

Some possible acid/base combinations to test:

- Clear soda pop (acid) and baking soda water (base).
- Clear soda pop (acid) and ammonia (base).
- Dilute clear window cleaner (base) with an acid like vinegar or clear soda pop.
- Lye [sodium hydroxide] (base) with an acid like vinegar or clear soda pop.
  (Lye is caustic—adult supervision is required.)

A chart is provided for the collection of data and a graph for plotting the data.

Experiment 8

Show Me the Starch!

Materials Needed

- tincture of iodine **[Iodine is VERY poisonous—DO NOT LET STUDENTS EAT any food items with iodine on them.]**
- a variety of raw foods, including:
  pasta
  bread
  celery
  potato
  banana (ripe) and other fruits
- 1 unripe (green) banana
- liquid laundry starch (or equal parts borax and corn starch mixed in water)
- absorbent white paper
- eye dropper
- cookie sheet
- marking pen
- knife

Objectives

In this experiment students will be introduced to "energy" molecules such as carbohydrates and starches that fuel our bodies.

The objectives of this lesson are for students to:

- Perform a simple test for starches.
- Use a control sample for comparison.

Experiment

I. Think About It

Read this section of the *Laboratory Notebook* with your students.

Ask questions such as the following to guide open inquiry.

- *Why do you think your body needs carbohydrates?*
- *What foods do you think contain carbohydrates? Why?*
- *What do you think carbohydrates are made of?*
- *Do you think all sugars are the same? Why or why not?*
- *Do you think there are different kinds of starches? Why or why not?*
- *Do you think grass would be good for you to eat? Why or why not?*

II. Experiment 5: Show Me the Starch!

Have the students read the entire experiment before writing an objective and a hypothesis.

Objective: Have the students write an objective (What will they be learning?). For example:

- *To determine which foods contain starch.*

Hypothesis: Have the students write a hypothesis. For example:

- *Potatoes contain starch. Celery does not.*

EXPERIMENT

❶ Have the students select food items to test and place them on a cookie sheet.

❷ Students will make a *control* by using an eye dropper to put a small amount of laundry starch (or a borax/cornstarch/water mixture) on a piece of absorbent paper and letting it dry before testing it.

❸ Have the students add a drop of iodine to the starch on the control paper and record the results in the chart provided in the *Results* section. [**Iodine is VERY poisonous — DO NOT LET STUDENTS EAT any food items with iodine on them.**]

❹ Have the students add a drop of iodine to each of the food samples and record their results.

❺-❻ Have the students compare the color of the *Control* to each of the food items tested and note which foods turned color.

Results

A chart is provided for recording the results of the experiment.

III. Conclusions

Have the students review the results they recorded for the experiment. Have them draw conclusions based on the data they collected.

IV. Why?

Read this section of the *Laboratory Notebook* with your students.
Discuss any questions that might come up.

V. Just For Fun

Students will observe how a banana changes as it ripens. As the banana ripens, the carbohydrates start to break apart, releasing small sugar molecules that give the banana its sweet taste.

Have the students begin with a green banana. Have them slice off a piece of the banana and test it with a drop of iodine. Again instruct them not to eat any food with iodine on it. Have them leave the banana at room temperature, and every third day have them test another slice of the banana. In the chart provided, have the students record their results after each test. When they have completed the experiment, have them review what happened over the course of the experiment.

Experiment 9

Mix It Up!

Materials Needed

- about 120 ml (1/2 cup) of each of the following:
  - water
  - ammonia
  - vegetable oil
  - rubbing alcohol
  - melted butter
  - vinegar
- small jars (7 or more)
- food coloring (6 colors)
- dish soap, 30 ml (2 tbsp.)
- eyedropper
- measuring cup and measuring spoons
- marking pen
- spoon
- ballpoint ink pens of various colors, including black*
- rubbing alcohol
- coffee filters, several (white)
- cardboard shoebox (or similar size box)
- tape
- scissors
- ruler

* **Pens:** Regular Bic® or other brand ballpoint pens can be used in this experiment. Black, blue, red, and green will give enough colors to compare. Also, multicolored ballpoint pens work well. Try to find one with at least 7 or 8 different colors.

Ballpoint pens work better than felt tip pens or markers. Buy inexpensive pens for this experiment because they will be taken apart.

Objectives

In this experiment students will explore homogeneous and heterogeneous mixtures.

The objectives of this lesson are for students to:

- Observe what happens when different substances are mixed together.
- Understand that a substance will mix or not mix with another substance depending on the characteristics of each substance.

Experiment

I. Think About It

Read this section of the *Laboratory Notebook* with your students.

Ask questions such as the following to guide open inquiry.

- *What substances can you think of that are mixtures?*
- *Do you think the foods we eat are mixtures? Why or why not?*
- *Do you think the air we breathe is a mixture? Why or why not?*
- *What do you think "dissolve" means?*
- *Do you think that any liquid that is stirred into another liquid will dissolve? Why or why not?*
- *Do you think there are some substances that will not stay mixed together? Why or why not?*

II. Experiment 9: Mix It Up!

Have the students read the entire experiment before writing an objective and a hypothesis.

Objective: Have the students write an objective (What will they be learning?). For example:

- *We will observe which solutions mix and which do not.*

Hypothesis: Have the students write a hypothesis. Some examples:

- *Oil will not dissolve in water without soap.*
- *Vegetable oil and butter will mix, but oil and water will not.*

If students need help coming up with a hypothesis, have them discuss some main points of mixtures—most substances are mixtures, like dissolves like, soap helps oil dissolve in water.

EXPERIMENT

If needed, other solutions can be substituted for the items listed. Try to pick at least two oily items and two water-based items.

Part I: What Mixes?

❶ In the *Results* section, a grid is provided for students to record their observations. They are to label the grid along the top and left side with the substances to be mixed and to mark the grid in a way that they don't make the same mixture more than once. To the right is an example.

	Water	Ammonia	Vegetable Oil	Rubbing Alcohol	Melted Butter	Vinegar
Water						
Ammonia						
Vegetable Oil						
Rubbing Alcohol						
Melted Butter						
Vinegar						

❷-❸ Have the students measure 60 ml (1/4 cup) of each liquid into its own jar, label each jar with its contents, and add a drop of food coloring to each.

❹ Based on the grid in the *Results* section, the students will select two liquids to mix together. Before combining the liquids, have the students predict whether or not the two will mix and discuss why they think this will happen. Using a clean jar for each test, have the students stir together 15 ml (1 tablespoon) of a liquid that has not been colored and 15 ml (1 tablespoon) of a colored liquid. In the grid in the *Results* section, have them record whether or not the two liquids mix. Was their prediction correct?

❺ Have them repeat Step ❹ for each combination of liquids.

Results (Part I)

A grid is provided for recording the results.

Part II: Soap, Oil, and Water

❶ Have the students measure 60 ml (1/4 cup) of water into a small glass jar and add one drop of food coloring.

❷-❸ Have them add 15 ml (1 tablespoon) of vegetable oil to the water, stir the water and oil, and record their results in the space provided in the *Results* section.

❹-❺ Have the students add 15 ml (1 tablespoon) of liquid dish soap to the oil/water mixture, stir thoroughly, and record their results.

❻ Have them add another 15 ml (1 tablespoon) of liquid dish soap to the mixture, stir thoroughly, and record their results.

III. Conclusions

Have the students review the results they recorded for the experiment. Have them draw conclusions based on the data they collected.

IV. Why?

Read this section of the *Laboratory Notebook* with your students.
Discuss any questions that might come up.

V. Just For Fun

Black Is Black?

In this experiment students will explore using paper chromatography as a technique for separating mixtures.

Have the students read the entire experiment before writing an objective and a hypothesis.

Objective: Have the students write an objective (What will they be learning?).

Hypothesis: Have the students state a hypothesis.

Experiment

❶ Have the students measure and pour 60 milliliters (1/4 cup) of alcohol into each of several small jars—one for each color pen to be tested. Have them label each jar with the color of ink to be tested in that jar.

❷-❸ Have them remove the thin plastic ink tube from inside each ballpoint pen and pull off the top or cut off the end of the tube.

❹ Have the students take an ink tube and swirl the open end in the alcohol. Some of the ink needs to be dissolved, but it is important not to let the color of the alcohol-ink mixture get too dark. A quick swirl with the open end of the pen is usually enough. Have them repeat for each ink tube using the jar labeled for that tube.

❺ Have the students cut the coffee filter paper into thin strips 6-12 mm (1/4 to 1/2 inch) wide and 13-15.5 cm (5 to 6 inches) long. They will need a strip for each known color of ink being tested and additional strips to test unknown colors.

6. One end of a paper strip will be placed in the dissolved ink in a jar, and the alcohol will be allowed to migrate upwards. The experiment works best if each strip is suspended in the alcohol without letting it touch the sides of the jars because the alcohol won't migrate past this point. To set up the experiment, students can lay a cardboard box on its side and tape the paper strips to the top of the box. The paper strips can then be suspended in the glass jars. Or students may come up with an idea for devising their own setup.

7. The colors in the ink will migrate up the absorbent paper strips. For the best results, students should leave the strips in the ink solutions overnight.

Results

Students are to let the paper strips dry and then tape them in the chart provided. Have them write down each original ink color and record the different colors each is made of.

Have them note which colors make up black. Depending on the brand of pen, the paper strip should show the black ink separated into all the colors or nearly all colors. Brown ink usually separates into all colors as well.

Most of the other color inks will be a mixture of fewer colors. Depending on the brand of pen, an ink color may also be a mixture of similar colors. For example, blue ink might be a mixture of only a dark blue and a light blue.

Unknown Ink

Now give the students an unknown sample to test. Make the unknown solution without letting the students observe. The unknown can contain inks from one pen or two different pens. If a mixture of two colors is to be used, first swirl each plastic tube in a separate jar of alcohol and then mix the two alcohol solutions together.

Suggest that the students give you an unknown.

Several unknowns can be set up. Again, let the paper strips soak in the solutions overnight.

Using the pattern of colors from the known pens that were tested, have the students try to identify which pen or pens were used for the unknowns.

Conclusions

Have the students review the results they recorded for the experiment. Have them draw valid conclusions based on the data they collected. Help them be specific. For example:

- *Black ink from Brand X is made of red, yellow, black, and brown.*
- *Red ink from Brand X is made of red and pink.*
- *Blue ink from Brand Y has only blue ink in it.*

Ask whether or not they proved or disproved their hypothesis.

Discuss those conclusions that are not valid. For example:

- *All black ink has yellow in it.* (This is not valid based on these data, because not all black inks were tested.)
- *Blue inks contain only blue.* (Again, not all blue inks were tested.)

Have the students record any sources of error. For example:

- *All of my alcohol evaporated and no dye went up the paper.*
- *The papers fell off the box and into the solution.*

Experiment 10

Testing for Lipids

Materials Needed

- 1-2 brown paper bags cut into about 20 squares, each 5 cm x 5 cm (2"x 2")
- wax paper, 8 pieces about 21.5 cm x 28 cm (8.5" x 11")
- paper towels
- tape
- knife
- scissors
- ruler
- marker or pen
- foods:

vegetable oil	margarine
butter	water
celery	orange
potatoes	cheese
banana	milk
avocado	cream
lard	

- several fat free and low fat food products, such as salad dressings and cheeses

Objectives

In this experiment students will test foods to determine whether they contain lipids.

The objectives of this lesson are for students to:

- Practice using controls in an experiment.
- Explore how simple tests can be used in scientific discovery.

Experiment

I. Think About It

Read this section of the *Laboratory Notebook* with your students.

Ask questions such as the following to guide open inquiry.

- *Do you think all foods contain lipids? Why or why not?*
- *Do you think it could be helpful to know whether a food contains lipids? Why or why not?*
- *Do you think lipids are important for a healthy diet? Why or why not?*
- *Do you think an experiment has to be complicated to be useful? Why or why not?*
- *Why do you think you would use a control in an experiment?*

II. Experiment 10: Testing for Lipids

Have the students read the entire experiment before writing an objective and a hypothesis.

Objective: Have the students write an objective. (What will they be learning?) Some examples:

- *To find out which foods contain lipids.*
- *We will see if a test can be done for lipids.*

Hypothesis: Have the students write a hypothesis. Some examples:

- *Vegetables do not contain lipids.*
- *Not all foods contain lipids.*

EXPERIMENT

Students will create a positive control and a negative control and then use these for comparison when determining whether a food contains or does not contain lipids.

Before beginning the experiment, have the students cut a brown paper bag into 13 squares 5 cm x 5 cm (2" x 2") in size. Foods may be substituted for any on the list that are not available. Students can also test additional foods of their choice.

❶ Have the students create a positive control and a negative control. The positive control will indicate the presence of lipids, and the negative control will indicate the absence of lipids.

Have students label one paper bag square "butter" and one "water." To make the positive control, they will place a small dab of butter on the paper bag square that has the "butter" label. For the negative control, they will place a drop of water on the "water" paper bag square. Have them allow both controls to sit for 10 minutes.

Have the students answer the questions in the *Laboratory Notebook.*

❷ After 10 minutes have the students wipe away any excess from the control papers and then hold each paper up against a bright light. Have them observe whether or not a translucent spot appears. The positive control should have a translucent spot and the negative control should not. Have them tape the controls in the boxes provided in the *Laboratory Notebook* and place a sheet of wax paper in front of and another behind the page.

❸ Have the students test the rest of the food items. Have them write the name of the food item on a paper bag square and then place a small dab or piece of food on the square.

❹ Have the students allow each sample to sit for 10 minutes. Any excess food should then be wiped off.

Results

❶ Have the students observe each sample and use their controls for comparison in determining whether the food contains lipids or does not contain lipids.

❷ Have them tape each sample in the correct column of the chart provided in the *Laboratory Notebook* and place a sheet of wax paper after each chart page.

III. Conclusions

Have the students review the results they recorded for the experiment. Have them draw conclusions based on the data they collected.

IV. Why?

Read this section of the *Laboratory Notebook* with your students.
Discuss any questions that might come up.

V. Just For Fun

Is It Really No Fat?

Students will test products that are labeled low fat or fat free to determine whether they contain lipids.

❶ Help the students gather the food items to be tested. They are instructed to choose ones that can be spread out on a paper bag square, for example, fat free and low fat salad dressings and fat free and low fat cheeses.

❷ Have the students cut a 5 cm x 5 cm (2″x 2″) brown paper bag square for each food item to be tested.

❸ Have them write the name of the food on a paper bag square, spread some of the food on the square, let it sit for 10 minutes, and then wipe off any excess food.

❹ Have them tape their samples in the appropriate column of the chart provided. Have them place wax paper pieces behind each the chart page.

Experiment 11

Gooey Glue

Materials Needed

- liquid laundry starch, 120 ml (1/2 cup)—or a mixture of 10 ml (2 tsp.) borax, 10 ml (2 tsp.) cornstarch, and about 320 ml (1 1/3 cup) water [equal parts cornstarch and borax mixed into enough water to dissolve them]
- Elmer's white glue, 60 ml (1/4 cup)
- Elmer's blue glue (or another glue different from white glue), 60 ml (1/4 cup)
- water
- 2 small jars
- marking pen that will write on glass
- Popsicle sticks for stirring
- measuring cup
- safety goggles
- rubber gloves
- apron
- 10 ml graduated cylinder
- beaker or glass jar
- glass stirring rod
- Nylon Synthesis and Rope Trick Kit from Home Science Tools
  Item # KT-ISNYLON
  http://www.hometrainingtools.com/

Objectives

In this experiment students will observe how chemical reactions can change the properties of polymers.

The objectives of this lesson are for students to:

- Explore how the properties of a polymer can change.
- Observe that different substances react in different ways when combined.

Experiment

I. Think About It

Read this section of the *Laboratory Notebook* with your students.

Ask questions such as the following to guide open inquiry.

- *What do you think a polymer is?*
- *Do you think polymers occur in nature or are man made? Why?*
- *What kinds of things do you think are made of polymers?*
- *Do you think the characteristics of a polymer can be changed? Why or why not?*
- *Do you think glue is a polymer?*
- *Do you think glue is always sticky? Do you think it can be changed so it is not sticky? Why or why not?*

II. Experiment 11: Gooey Glue

Have the students read the entire experiment before writing an objective and a hypothesis.

Objective: Have the students write an objective (What will they be learning?). For example:

- *In this experiment we will observe a change in properties as two polymers are mixed together.*

Hypothesis: Have the students write a hypothesis. Some examples:

- *Liquid starch will change the polymer properties of white Elmer's glue.*
- *The other glue will change in the same ways as white Elmer's glue when liquid starch is added to it.*

EXPERIMENT

Part I

In this experiment the students will combine Elmer's white glue with liquid laundry starch (or a borax-cornstarch mixture). The liquid laundry starch will change the properties of the glue and create a soft, malleable ball something like Silly Putty.

❶ Have the students put a small amount of Elmer's white glue on their fingertips and observe the color and consistency of the glue. Have them record their observations in the *Results* section.

❷ Have the students observe the properties of the liquid starch (or borax-cornstarch mixture) by pouring a small amount on their fingers or in a jar and noting the color and consistency. Have them record their observations.

Because Elmer's glue is difficult to measure, Steps ❸-❻ provide a way to measure a given amount of glue directly into the jar. The amount of starch added is not that important, but this is given as a guide. The length of time the glue stays in the starch is more critical—more time in the starch results in a stiffer glue-starch ball.

❸-❹ Have them measure and pour 60 ml (1/4 cup) of water into one of the jars and use a marker to draw a small line at the water level.

❺ Then have them add another 60 ml (1/4 cup) of water to the jar and use a marker to draw a small line at the new water level.

❻ Have them pour the water out.

❼ Have them fill the jar to the first mark with Elmer's glue.

❽ Have them add liquid starch (or borax-cornstarch mixture) to the second line.

❾ Have the students mix the glue and starch with a Popsicle stick and record any changes in consistency and color. The consistency of the glue will immediately change upon addition of the laundry starch, but it will still be sticky. The laundry starch has to be kneaded into the glue.

❿ Once enough laundry starch has been mixed into the glue for the mixture to begin to get firm, students can remove the glue-starch ball from the jar and knead it with their fingers.

Have them observe the consistency and color and record their results.

Have the students note the change in the properties of the glue, and help them think of ways to accurately describe the texture of the glue-starch ball. Some examples: bouncy, stretchy, somewhat elastic like a rubber band, blue/white in color. (Answers may vary.)

The mixture continues to get harder as the glue is allowed to react with the starch. You can have the students make more than one glue-starch ball and allow one to be mixed for a longer time for comparison.

Results

Space is provided for recording observations.

Part II

Have the students repeat the experiment using a different kind of glue. Elmer's blue glue works best, but any other brand of nonwhite glue can be used. The blue glue does not react in exactly the same manner as the white glue, and it will have a different consistency. Have the students predict whether there will be different results when liquid laundry starch is added to the blue glue than when it was added to the white glue.

❶ Have the students follow Steps ❸-❻ in **Part I** of this experiment.

❷-❹ This time they will fill the jar to the first mark with the Elmer's blue glue or another glue that is different from the white glue, then add liquid starch to the second mark and mix.

❺ Have the students record their observations.

III. Conclusions

Have the students review the results they recorded for the experiment. Have them draw conclusions based on the data they collected. Help the students relate the changes of the glue to the information given in the text.

Some conclusions can be inferred from statements that are based on the data collected. For example, because the two glues behaved differently, it is possible that the chemical composition of each is different. Also, they have a different color and texture. However, this experiment did NOT prove that they are different chemically even though the experimental data suggest that this might be true. Discuss the need for additional investigation in order to reach valid conclusions about the chemical composition of the two types of glue.

IV. Why?

Read this section of the *Laboratory Notebook* with your students.
Discuss any questions that might come up.

V. Just For Fun

Make nylon!

In this experiment students will mix two liquids that will react to form a nylon polymer.

1. Have the students look up the chemistry for Nylon 6-10 online or at the library. In the space provided, have them record what they discover.

2. Have the students get out the Nylon Synthesis Kit and read the instructions. Have them gather the equipment needed for the experiment. It is recommended that they wear safety goggles, rubber gloves, and an apron during the experiment.

   After they have performed the experiment, have them record their observations and conclusions in the space provided.

Experiment 12

Amylase Action

Materials Needed

- tincture of iodine [VERY POISONOUS—DO NOT LET STUDENTS EAT any food items that have iodine on them]
- bread (1-2 slices)
- timer
- wax paper
- marking pen
- cup
- one raw egg
- one raw onion
- table salt
- clear liquid dish washing detergent
- rubbing alcohol (isopropanol, 70-90%)
- wooden stir stick or Q-tip
- coffee filter (any color)
- sieve
- 2 glass jars or large test tubes
- measuring cup and measuring spoons
- blender

Objectives

In this experiment students will observe how an enzyme breaks down a protein polymer.

The objectives of this lesson are for students to:

- Perform a simple test using a chemical reaction.
- Observe how saliva begins food digestion by breaking down starch.

Experiment

I. Think About It

Read this section of the *Laboratory Notebook* with your students.

Ask questions such as the following to guide open inquiry.

- *What do you think proteins are made of?*
- *Do you think proteins can be changed? Why or why not?*
- *What kinds of jobs do you think molecular machines do?*
- *What do you think is the function of DNA in the body?*
- *Do you think a protein can be any shape it wants to be? Why or why not?*

II. Experiment 12: Amylase Action

As a review before starting this experiment, ask questions such as the following. Students can refer to the textbook to find the answers.

- *What molecule have we studied that is in bread?*
  starch (amylose)
- *What happens to food when we put it in our mouth?*
  It begins to be broken down (digested).
- *What do you think happens to amylose in our mouth?*
  It begins to break down (is digested).
- *Why do you think iodine is used in this experiment?*
  To show that starch has been broken down into smaller molecules (sugars).

- *How?*
  Saliva has an enzyme, amylase, that breaks down amylose. Digestion of food begins in the mouth. Iodine will react with starch but not with the sugar molecules that the starch polymer chain is made of.

Have the students read the entire experiment before writing an objective and a hypothesis.

Objective: Have the students write an objective (What will they be learning?). For example:

- *We will investigate the cutting action of proteins in saliva.*

Hypothesis: Have the students write a hypothesis. For example:

- *Saliva will cut the starch molecules in bread, and iodine will react differently when placed on unchewed bread than on chewed bread.*

EXPERIMENT

❶-❷ Have the students break the bread into several small pieces and then chew one piece for 30 seconds, another piece for 1 minute, and a third piece for as long as possible (several minutes).

❸ Each time, after chewing the bread, have them spit it onto a piece of wax paper and label the wax paper with the length of time the bread has been chewed.

❹-❺ Have the students take three small pieces of unchewed bread, place one next to each of the chewed pieces, and add a drop of iodine to each piece of bread—chewed and unchewed. [Iodine is VERY POISONOUS—DO NOT LET STUDENTS EAT any food items that have iodine on them]

❻ Have them record their observations in the chart in the *Results* section.

❼ Have the students take two more small pieces of bread. Have them spit into a cup several times to collect as much saliva as possible. Then they will soak both pieces of bread in the saliva, place one piece in the refrigerator, leave the other piece at room temperature, and let all the samples soak for 30 minutes.

❽ After 30 minutes have the students add a drop of iodine to each sample and record their results. They should have observed a decrease in the black color of the iodine with the bread that has been chewed for longer times. Also, the refrigerated bread should be darker black than the unrefrigerated bread because the iodine/starch reaction is slowed by the cooler temperature.

Results

Charts are provided for recording observations.

III. Conclusions

Have the students review the results they recorded for the experiment and summarize the data. Discuss with them what is likely to have occurred:

1) The chewed bread and the bread with saliva at room temperature showed a decrease in the black color after iodine was added.
2) We know that bread contains starch (amylose).
3) We know that saliva has protein machines in it that begin the digestion of the food.
4) The color change that happens with the addition of iodine shows that it is likely that saliva contains a protein machine that breaks down the amylose in bread.

These four statements are based both on data that has actually been collected (1) and information that has been gathered from other sources (2 and 3). The conclusion (4) is a likely conclusion and most probably correct, but further data would need to be collected to prove that the concluding statement is true. Discuss how this investigation could lead to other experiments to prove the conclusion. For example, a protein analysis could be performed to determine which proteins are in saliva.

This is how real science works. Initial observations lead to additional experiments, which hopefully yield enough data to prove or disprove a given statement. Sometimes it takes years to collect all of the data, and sometimes enough data is never gathered. In the latter case, the hypothesis remains unproven, and the most valid conclusions are only likely ones.

Have the students write valid conclusions based on the data they have collected.

IV. Why?

Read this section of the *Laboratory Notebook* with your students.
Discuss any questions that might come up.

V. Just For Fun

Which Has More DNA? An Onion or an Egg?

This experiment is a bit more involved than most the students have already done. They will be extracting actual DNA from an onion and an egg. They will compare the samples to see which has more DNA. Each step in this experiment has an explanation about what is being done to extract nucleic acids from living tissues.

Guide open inquiry with questions such as:

- *Do you think you can do an experiment where you will extract DNA from plant or animal cells? Why or why not?*
- *Do you think you can see DNA? Why or why not?*
- *Do you think plants and animals have similar DNA or do they have totally different DNA? Why?*
- *Do you think an egg has more DNA than an onion or does the onion have more DNA? Why?*

The following experimental descriptions and steps are written as they appear in the Laboratory Notebook.

Hints for performing this experiment: Make your measurements as accurate as you can. Try not to lose too much of the samples as you transfer them from container to container.

Materials

one raw egg
one raw onion
table salt
clear liquid dish washing detergent
rubbing alcohol (isopropanol, 70-90%)
wooden stir stick or Q-tip
coffee filter (any color)
sieve
2 glass jars or large test tubes
measuring cup and measuring spoons
blender

❶ Lysing the cells in the sample

Recall that living tissues are made of cells and the nucleic acids are inside the cells. In order to extract nucleic acids, the cells need to be opened, or *lysed*. The first step of this experiment opens, or lyses, the cells. This is accomplished by using a combination of detergents and enzymes.

Part A: Prepare the sample for lysis

1) Put a shelled raw egg in a blender and add 240 ml (1 cup) cold water. To this mixture, add 5 ml (1 teaspoon) of table salt and blend the sample on high speed until it is pureed (about 15-20 seconds).

2) Pour the egg sample through a sieve into a glass jar or large test tube.

3) Repeat for the onion but this time add .5 liter (2 cups) of water.

4) You need to have at least 60 ml (1/4 cup) of liquid for each sample before continuing with the experiment.

Part B: Lysis

1) Add detergent to the mixtures to break the cells open. For each sample, use 15 ml (1 tablespoon) of detergent per cup of cell mixture.

2) Gently swirl the cell, water, and detergent mixture by rotating the container or mixing with a stick or spoon, being careful not to create foam.

3) Allow the mixture to sit for 5 minutes, gently swirling intermittently.

❷ Separating the DNA from the cell material

Once the cells are lysed, or broken open, you'll have a mixture of DNA, RNA, proteins, and other cell parts. The DNA and RNA need to be separated from the rest of the mixture. This is accomplished with the use of alcohol. The nucleic acids are not soluble in alcohol; therefore, they will precipitate out of the solution.

1) Tilt the jar or test tube and slowly add 60 ml (1/4 cup) of isopropanol per 60 ml (1/4 cup) of the cell/water mixture, carefully pouring it down the inside of the jar or test tube. This must be done slowly without agitating the mixture. The alcohol will float to the top of the jar or test tube, and the DNA will precipitate at the water-alcohol interface (the area where the alcohol and water meet).

2) Perform Step 1) for both samples.

❸ Pulling out the DNA

After the DNA is separated from the cell parts, it can be extracted, or pulled out of the solution. This is accomplished with the use of a wooden stick or Q-tip. Although all nucleic acids can be removed from the cells, only DNA survives the procedure. RNA is chewed up, or degraded, by enzymes during the process. DNA is more robust than RNA and is not easily degraded by enzymes.

1) Take the Q-tip or wooden stick and insert it into the alcohol layer. Gently touch the alcohol-water interface, and gently swirl the stick or Q-tip, pulling up slightly. The DNA will collect on the stick or Q-tip and long strands should be visible.

2) Continue spinning and collecting the DNA for a few seconds.

3) Pull the stick out and place the DNA on a coffee filter to dry.

4) Write the results and your conclusions in the spaces provided at the end of the experiment.

Note: If this experiment is successful, students should find that they get DNA from the onion but not from the egg. An egg is made of just one cell, so there will not be enough DNA to be extracted.

Commentary

The protocol, or procedure to be followed, in this experiment describes a simple method for extracting DNA from living tissues.

The steps in this experiment can be adapted to use with a variety of different samples by adjusting the volumes for the sample being prepared. If you are lysing animal cells, add 5 ml (1 teaspoon) of meat tenderizer. The enzymes in meat tenderizer are needed to open animal cells.

Troubleshooting

Troubleshooting is part of doing science. Few experiments work the first time. Many new discoveries are made by scientists when their experiments "fail."

Frequently Asked Questions

Troubleshooting is part of doing science. Few experiments work the first time. Many new discoveries are made by scientists when their experiments "fail."

Frequently Asked Questions

Q. If I do not get at least 60 ml (1/4 cup) of liquid from Step ❶, Part A, should I continue?

A. No. If you do not get at least 60 ml (1/4 cup) of material, blend the sample again, adding more water. With smaller volumes, there may not be enough DNA *extracted* to be visible.

Q. What if I do not see an alcohol-water interface?

A. If you do not see an alcohol-water interface, try adding more alcohol, being careful not to agitate the sample. If you still do not see an interface, check the concentration of your alcohol, and make sure it is not less than 70% alcohol. *If your alcohol is 70% or more and you still do not see an interface, add twice the volume of alcohol to the sample.* You should eventually see an interface. If this fails, discard the sample and start over, making sure you use the amount of water specified in Step ❶.

Q. What if I see foam?

A. Carefully remove the foam with an eyedropper *without agitating the sample.*

Q. What if I do not get any DNA?

A. There could be several reasons you do not see any DNA.

1) You did not use enough starting material. Repeat the experiment and double the amount of starting material.

2) You did not use enough detergent. Repeat the experiment using more detergent.

3) You did not use the right kind of detergent. Repeat the experiment with a different detergent.

4) You did not let the sample sit long enough to break open the cells. Repeat the experiment and allow the sample to sit for a longer period of time.

5) You did not add detergent or enzymes to the sample. Repeat the experiment adding enzyme or detergent or both.

6) You did not add enough alcohol. Add more alcohol.

7) The alcohol you added was not concentrated enough. Add 70-90% rubbing alcohol.

8) There is not enough salt in the water mixture to precipitate the sample. Add 5 ml (1 teaspoon) of table salt to your water-cell-alcohol mixture. Swirl. Add more rubbing alcohol until you see an interface, then try to pull out the DNA.

Note: This experiment can be repeated using samples of other living things including but not limited to:

- Vegetable tissue such as spinach, peas, green beans, broccoli, onions, etc.
- Grains such as wheat germ, corn, oatmeal, seeds, or yeast.
- Animal tissue such as eggs, chicken or beef livers, chicken hearts, etc.

All items must be *uncooked.*

More REAL SCIENCE-4-KIDS Books
by Rebecca W. Keller, PhD

Building Blocks Series yearlong study program — each Student Textbook has accompanying Laboratory Notebook, Teacher's Manual, Lesson Plan, Study Notebook, Quizzes, and Graphics Package

Exploring Science Book K (Activity Book)
Exploring Science Book 1
Exploring Science Book 2
Exploring Science Book 3
Exploring Science Book 4
Exploring Science Book 5
Exploring Science Book 6
Exploring Science Book 7
Exploring Science Book 8

Focus On Series unit study program — each title has a Student Textbook with accompanying Laboratory Notebook, Teacher's Manual, Lesson Plan, Study Notebook, Quizzes, and Graphics Package

Focus On Elementary Chemistry
Focus On Elementary Biology
Focus On Elementary Physics
Focus On Elementary Geology
Focus On Elementary Astronomy

Focus On Middle School Chemistry
Focus On Middle School Biology
Focus On Middle School Physics
Focus On Middle School Geology
Focus On Middle School Astronomy

Focus On High School Chemistry

Super Simple Science Experiments

21 Super Simple Chemistry Experiments
21 Super Simple Biology Experiments
21 Super Simple Physics Experiments
21 Super Simple Geology Experiments
21 Super Simple Astronomy Experiments
101 Super Simple Science Experiments

Note: A few titles may still be in production.

Gravitas Publications Inc.
www.gravitaspublications.com
www.realscience4kids.com